NORTH AMERICAN MAMMALS

A Photographic Album for Artists and Designers

Selected and Edited by

James Spero

Dover Publications, Inc.

New York

For Mimi (*Felis horribilis* Wilson)

ACKNOWLEDGMENTS

With the exception of the illustrations listed below, all of the photographs in this book came from the collection of the Office of Audio Visual, Fish and Wildlife Service, U.S. Department of the Interior, Washington, D.C. The publisher would like to thank Ms. Bea Boone and Mr. Robert Hines of the Service for their aid. The photographs of the polar bear, as well as the wolf on page 91, were taken by Mr. Edmund V. Gillon, Jr., especially for this publication. Mr. Robert M. Zewadski, Technical Editor, Illinois Natural History Survey, was kind enough to supply the photographs of the chipmunk, southeastern shrew, mink, meadow mouse (page 48), pine mouse and long-tailed weasel. The photographs of the fisher, walrus and wolverine came from the San Diego Zoo. Glossy prints are available, at a fee, from the zoo. For a price schedule and other particulars, please write to: San Diego Zoo Photo Lab, P.O. Box 551, San Diego, Ca. 92112.

Published in Canada by General Publishing Company, Ltd.,
30 Lesmill Road, Don Mills, Toronto, Ontario.
Published in the United Kingdom by Constable and Company,
Ltd., 10 Orange Street, London WC2H 7EG.

North American Mammals: A Photographic Album for Artists and Designers is a new work, first published by Dover Publications, Inc., in 1978.

DOVER *Pictorial Archive* SERIES

North American Mammals: A Photographic Album for Artists and Designers belongs to the Dover Pictorial Archive Series. Up to ten illustrations may be used on any one project or in any single publication, free and without special permission. Wherever possible, include a credit line indicating the title of this book, photographer and publisher. Please address the publisher for permission to make more extensive use of illustrations than that authorized above.

International Standard Book Number: 0-486-23601-3
Library of Congress Catalog Card Number: 77-88736

Manufactured in the United States of America
Dover Publications, Inc.
180 Varick Street
New York, N.Y. 10014

Frontispiece: Bobcat (Bay Lynx). *Felis rufa.* (Charles L. Cadieux)

CONTENTS

Armadillo. *Dasypus novemcinctus mexicanus.* (Luther C. Goldman)

Badger. *Taxidea taxus*. (Luther C. Goldman)

Badger. *Taxidea taxus.* (E. P. Haddon)

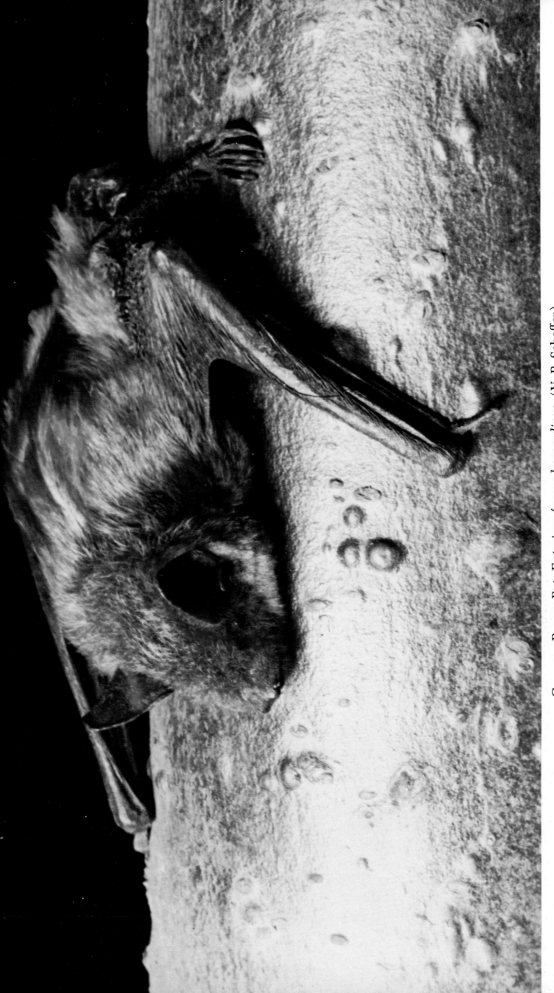

Common Brown Bat. *Eptesicus fuscus bernardinus.* (V. B. Scheffer)

Common Brown Bat. *Eptesicus fuscus bernardinus.* (V. B. Scheffer)

Black Bear. *Ursus americanus.* (E. P. Haddon)

Black Bear. *Ursus americanus.* (E. P. Haddon)

Grizzly Bear. *Ursus horribilis.* (National Park Service)

Polar Bear. *Thalarctos maritimus.* (Edmund V. Gillon, Jr.)

Polar Bear. *Thalarctos maritimus.* (Edmund V. Gillon, Jr.)

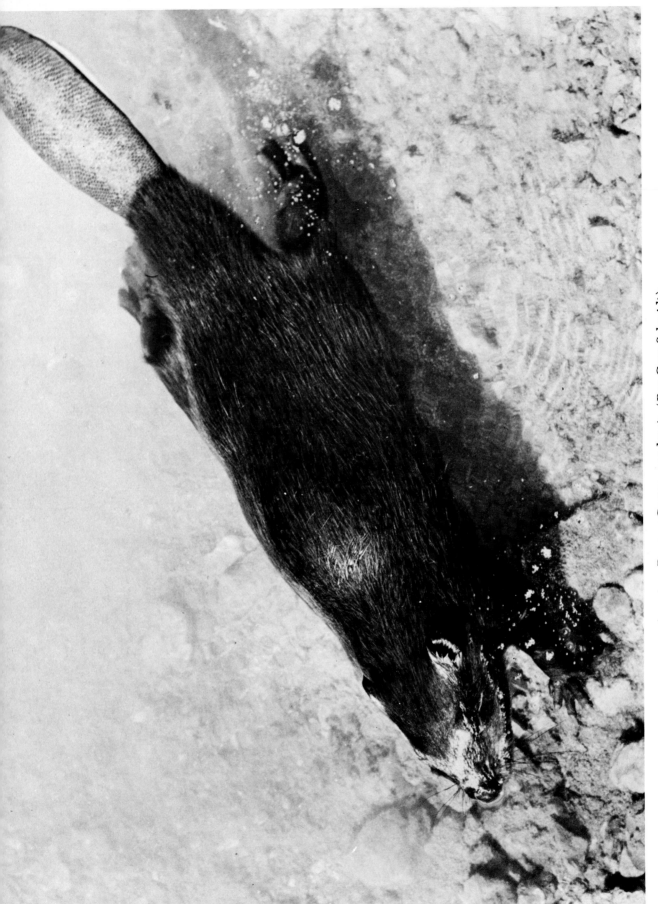

American Beaver. *Castor canadensis.* (Rex Gary Schmidt)

Mountain Beaver. *Aplodontia rufa.* (V. B. Scheffer)

Bison (Buffalo). *Bison bison.* (Rex Gary Schmidt)

Bison (Buffalo). *Bison bison.* (E. P. Haddon)

Bobcat (Bay Lynx). *Felis rufa.* (B. E. Foster)

Bobcat (Bay Lynx). *Felis rufa.* (L. J. Goldman)

Cacomistle (Ringtail, Civet, Civet Cat). *Bassariscus astutus.* (W. M. Rush)

Eastern Chipmunk. *Tamias striatus*. (Karl H. Maslowski)

Coyote. *Canis latrans.* (E. R. Kalmbach)

Coyote. *Canis latrans.* (E. R. Kalmbach)

Coyote. *Canis latrans.* (E. P. Haddon)

Key Deer. *Odocoileus virginianus clavium.* (Richard Thompson)

Mule Deer. *Odocoileus hemionus hemionus.* (E. P. Haddon)

Mule Deer. *Odocoileus hemionus hemionus.* (E. P. Haddon)

White-tailed Deer. *Odocoileus virginianus.* (E. P. Haddon)

White-tailed Deer. *Odocoileus virginianus.* (E. P. Haddon)

Black-footed Ferret. *Mustela nigripes.* (Luther C. Goldman)

Black-footed Ferret. *Mustela nigripes.* (Luther C. Goldman)

Fisher. *Martes pennanti.*

Gray Fox. *Urocyon cinereoargenteus.* (Allen M. Pearson)

Red Fox. *Vulpes fulva.* (E. P. Haddon)

Pocket Gopher. *Geomys areuarius.*

Varying Hare. *Lepus americanus.* (V. B. Scheffer)

Jackrabbit. *Lepus californicus.* (W. P. Taylor)

Mink. *Mustela vison.* (Maslowski and Goodpaster)

Pacific Mole. *Scapanus orarius.* (V. B. Scheffer)

Moose. *Alces alces gigas.* (J. Malcolm Greany)

Moose. *Alces alces gigas.* (J. Malcolm Greany)

Rocky Mountain Goat. *Oreamnos americanus.* (David L. Spencer)

Meadow Mouse (Meadow Vole). *Microtus pennsylvanicus.* (O. J. Murie)

Meadow Mouse (Meadow Vole). *Microtus pennsylvanicus.* (Karl H. Maslowski)

Pine Mouse (Pine Vole). *Pitymys pinetorum*. (Karl H. Maslowski)

49

Musk Ox. *Ovibos moschatus.*

Musk Ox. *Ovibos moschatus.* (Cecil Rhode)

Muskrat. *Ondatra zibethicus.* (V. B. Scheffer)

Ocelot. *Felis pardalis.* (Charles E. Most)

Opossum. *Didelphis virginiana.* (Frank M. Blake)

Opossum. *Didelphis virginiana.* (Frank M. Blake)

Otter. *Lutra canadensis.* (V. B. Scheffer)

Sea Otter. *Enhydra lutris lutris.* (Karl W. Kenyon)

Collared Peccary. *Tayassu tajacu*. (Luther C. Goldman).

Pika. *Ochotona princeps.* (E. R. Warren)

Porcupine. *Erithizon dorsatum.* (D. A. Spencer)

Porcupine. *Erithizon dorsatum.* (I. K. Couch)

Yellow-haired Porcupine. *Erithizon dorsatum epixanthum.* (Ward M. Sharp)

Black-tailed Prairie Dog. *Cynomys ludovicianus.* (E. R. Kalmbach)

Black-tailed Prairie Dog. *Cynomys ludovicianus.* (E. R. Kalmbach)

Pronghorn (Pronghorn Antelope). *Antilocapra americana.* (E. P. Haddon)

Puma (Mountain Lion, Cougar). *Felis concolor.* (Colorado Game and Fish)

Pygmy Rabbit. *Sylvilagus idahoensis.* (O. J. White)

Raccoon. *Procyon lotor.* (Rex Gary Schmidt)

Raccoon. *Procyon lotor.* (Rex Gary Schmidt)

Merriam's Kangaroo Rat. *Dipodomys merriami.* (E. R. Kalmbach)

Norway Rat (Brown Rat). *Rattus norvegicus*. (W. D. Fitzwater)

Northern Fur Seal. *Callorhinus ursinus.* (V. B. Scheffer)

Monk Seal. *Monachus albiventer.* (Eugene Kridler)

Steller's Sea Lion. *Eumetopias jubata.* (LeRoy W. Sowl)

Bighorn Sheep. *Ovis canadensis.* (Charles G. Hansen)

Bighorn Sheep. *Ovis canadensis.* (A. Schlechten)

Bighorn Sheep. *Ovis canadensis.* (Joe Mazzoni)

Southeastern Shrew. *Sorex longirostris longirostris* Bachman. (Illinois Natural History Survey)

Striped Skunk. *Mephitis mephitis.* (Rex Gary Schmidt)

Striped Skunk. *Mephitis mephitis.* (V. B. Scheffer)

Gray Squirrel. *Sciurus carolinensis.* (E. P. Haddon)

Del Marva Fox Squirrel. *Sciurus niger cinereus.* (William H. Julian)

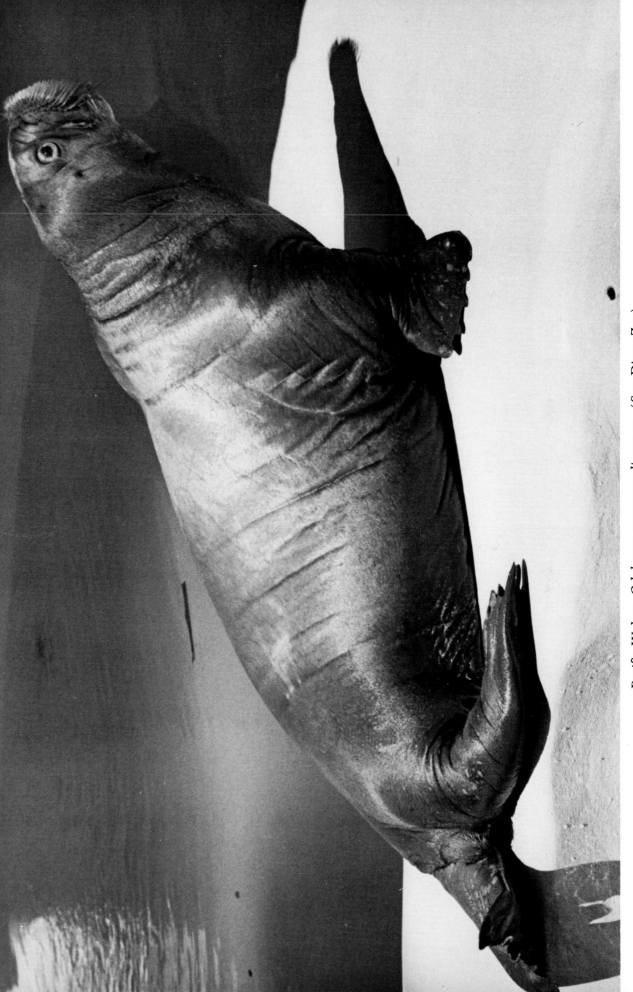

Pacific Walrus. *Odobenus rosmarus divergens.* (San Diego Zoo)

Wapiti (Elk). *Cervus canadensis nelsoni*. (E. P. Haddon)

Wapiti (Elk). *Cervus canadensis nelsoni.* (C. J. Henry)

Long-tailed Weasel. *Mustela frenata*. (Maslowski and Goodpaster)

Gray Wolf (Timber Wolf). *Canis lupus*. (Edmund V. Gillon, Jr.)

Gray Wolf (Timber Wolf). *Canis lupus.* (Don Reilly)

Texas Red Wolf. *Canis niger rufus.* (Hal Swiggett)

Wolverine. *Gulo gulo luscus.* (San Diego Zoo)

Woodchuck (Groundhog). *Marmota monax.* (Robert W. Hines)

Woodchuck (Groundhog). *Marmota monax.* (W. M. Kent)